Investigations in Number, Data, and Space®

Assessment Sourcebook

Grade 1

Scott
Foresman

Editorial Offices: Glenview, Illinois • Parsippany, New Jersey • New York, New York
Sales Offices: Parsippany, New Jersey • Duluth, Georgia • Glenview, Illinois
Carrollton, Texas • Ontario, California
http://www.scottforesman.com

The End-of-Unit Assessment Tasks included in each grade level's *Assessment Sourcebook* were developed by experienced classroom teachers in conjunction with the authors of the *Investigations* curriculum. These teachers have brought years of teaching experience, an in-depth knowledge of the *Investigations* curriculum, and a deep understanding of how children learn mathematics to the development of these assessment materials.

TERC Staff

Karen Economopoulos
Lucy Wittenberg
Katie Bloomfield

Megan Murray
Lorraine Brooks

Collaborating Teachers

Grade 1 Rose Christiansen   Malia Scott

Grade 2 Jennifer DiBrienza   Elizabeth Sweeney

Grade 3 Katie Bloomfield   Nancy Horowitz

Grade 4 Nancy Buell   Jan Rook

Grade 5 Katherine Casey   Gary Shevell

This project was supported, in part,
by the
National Science Foundation
Opinions expressed are those of the authors
and not necessarily those of the Foundation

TERC

The *Investigations* curriculum was developed at TERC (formerly Technical Education Research Centers) in collaboration with Kent State University and the State University of New York at Buffalo. The work was supported in part by National Science Foundation Grant No. ESI-9050210. TERC is a nonprofit company working to improve mathematics and science education. TERC is located at 2067 Massachusetts Avenue, Cambridge, MA 02140.

Contents

Introduction

Assessment plays a critical role in teaching and in learning. The *Assessment Sourcebooks,* a new resource designed to accompany the *Investigations in Number, Data, and Space* curriculum units in grades 1–5, complement and support the assessment component of *Investigations* by offering teachers further opportunities to gather information about students' growing mathematical understanding.

Each *Sourcebook* provides teachers with sets of Assessment Tasks designed to assess students' understanding of the most important mathematical ideas covered in that grade level's curriculum units. Each *Sourcebook* also provides information for teachers about the mathematical significance of each assessment task; suggestions on how to observe students and evaluate their work; and unit checklists of mathematical emphases.

For you, the Teacher

Each **End-of-Unit Task** is accompanied by an explanation of the mathematical content it addresses as well as short descriptions of the types of student responses a teacher might expect. A reproduction of each **Assessment Master** needed for the task is shown for your reference.

Each assessment task includes a **What to Look For ...** bulleted list of suggestions to guide you as you examine student work. This information is designed to help you assess the work of each student as well as the work of your class as a whole. Each task details connections to related work in the unit and to relevant Teacher Notes and Dialogue Boxes, when appropriate, offering you additional support as you examine student work.

While the introduction to each task or set of tasks usually refers to the class as a whole, the bulleted list of suggestions refers to the individual student. Rather than using the often awkward *he or she, his or her*, and so on, we have chosen to alternate between masculine and feminine pronouns: one list uses the masculine, and the next list uses the feminine. We hope that this device will make each list seem more personal. We also hope that referring to the class in the introduction and to the individual student in the list will help you remember to strike a balance between the two in your assessment program.

Occasionally a task may require extra materials, such as manipulatives or graph paper, or the task may require additional information, such as how to adjust the task for a student who may need more or less challenge. Such information appears under **Notes** in the left margin of the text.

Also included in the left margin of the text is space labeled **Your Notes** for you to make relevant notes about either the task or your students' responses to the task. These notes can be useful as you examine student work and as you assign and evaluate each set of tasks in future years.

Each set of End-of-Unit Assessment Tasks is followed by a list of the important Mathematical Emphases that are covered throughout that unit. This list is also provided as a **Checklist of Mathematical Emphases** master, with space to record information about three students at a time. (The checklist can be reproduced as often as need be to provide one for every three students.) These checklists provide a concise place to make short notes about individual students.

How to Use These Assessments
Each Unit Assessment is designed to be used, in its entirety, at the end of a curriculum unit. The assessments follow the suggested sequence of units at each grade level. If the assessment tasks are done out of sequence, students may not have the necessary mathematical experiences to match the expectations. These assessments should be used in addition to other assessments presented in the unit and along with the samples of student work that you have chosen to save (see **Choosing Student Work to Save,** which appears toward the end of each curriculum-unit book). This collection offers a picture of a student's understanding of the mathematical concepts and skills presented in the unit.

For Your Students

Each Assessment Task requires one or more **Assessment Masters,** which are intended to be used as student recording sheets. The masters are provided in both English and Spanish and are meant to be copied for each student in your class. In some situations it may be necessary for students to use the back side of a sheet or to continue their work on another sheet of paper. While most tasks require only the appropriate assessment masters, some may require additional materials, such as manipulatives, graph paper, or measuring tools that students have been using throughout a particular unit.

End-of-Unit Assessment Tasks

Tasks 1A and 1B

In this unit, students have been counting in a variety of ways. The assessment tasks outlined on Assessment Master 1 will give you an opportunity to see students count a quantity up to twenty. It is expected that students are able to keep track of how many objects they have counted so far and how many more objects are yet to be counted. They should be able to count accurately up to twenty. You may want to refer to your notes from the "Counting 20" Teacher Checkpoint on pages 17 and 18 in Mathematical Thinking at Grade 1.

Name _____ Date _____

Assessment Master 1

End-of-Unit Assessment Tasks

1A. Draw 14 circles in the space below.

1B. How many more circles do you need to draw to have 20 in all? Add these to the circles you have drawn.

Your Notes

What to look for ...

- Can the student count accurately? Does the student say the number names in sequence? Does she say one number word for each circle? Does she skip any circles? Does she count any circles twice?

- Is she able to keep track of how many circles she has counted so far?

- Does the student have a strategy for finding out how many more circles she needs to draw (in this case, to make twenty)? Has she added 6 circles to the 14 circles?

Task 2

NOTE

If students are having difficulty working with combinations of nine, you might suggest that they solve the problem using a smaller total, such as seven or five.

As your students perform the task outlined on Assessment Master 2, you can get a sense of how they think about number combinations, how they solve complex problems, and how they keep track of and record their work. It is expected that students will be able to find at least one combination of circles and squares, and it is likely that many first graders will be able to find more than one combination. At this point in grade 1, many students are likely to record their results using a combination of pictures and numbers, although some students may also record their results using equations. The "Exploring Multiple Solutions" Teacher Note on page 49 of Mathematical Thinking at Grade 1 offers information about examining student work.

Name _____   Date _____

Assessment Master 2

End-of-Unit Assessment Task

2. You have 9 shapes. Some of them are circles, and some of them are squares. How many of each shape could you have? Find as many different solutions as you can.

YOUR NOTES

_____

_____

_____

_____

What to look for ...

- What strategies does the student have for approaching the problem? Does he:
  - ➤ choose to model the problem with cubes or counters?
  - ➤ draw pictures?
  - ➤ mentally know the combinations of 9 and record (some of) them using equations?

- If the student uses counters, does he randomly take counters of two different colors, count them, and then adjust the numbers so that he has nine in all? Or does he seem to have a systematic plan for making nine by taking away a quantity from one group and then adding on the same quantity to the other group?

- Can the student easily keep track of the total number of circles and squares as well as the individual number of each?

- How does the student record his solutions on paper? Does he draw pictures of circles and squares? Draw other symbols to represent the counters he used to model the problem? Write numbers? Write equations? Write words? Does the student label how many "things" are circles and how many are squares?

- Is the student able to think of multiple solutions to the problem? If so, does it appear that his solutions are random? Did the student find pairs of "opposite" solutions (for example, 1 circle and 8 squares; 8 circles and 1 square)? Did the student create ordered lists?

Task 3

As students solve the problem outlined on Assessment Master 3, you have an opportunity to observe how they represent information about a series of groups and to discover what strategies they use for combining several quantities. This task is similar to the activity students did when you read them Rooster's Off to See the World (or a similar book).

NOTE

Students should be able to accurately represent each quantity and determine the total. Some students will count each animal individually and others may combine smaller numbers to compute the total: 2 + 2 = 4; 4 + 4 = 8; 8 + 3 = 11.

Name _____   Date _____

Assessment Master 3

End-of-Unit Assessment Task

3.   At the zoo, the children saw many animals.
There were 2 monkeys on a tree,
4 tigers on the ground,
3 lions in a den,
and 2 zebras inside a fenced area.
How many animals did the children see?

Your Notes

_____

_____

_____

_____

What to look for ...

- Is the student's answer correct?

- Does the student have a way to represent the number of animals in each subgroup? Does she use numbers or number words to show how many? Does she draw pictures? Make tally marks?

- Does the student use concrete materials, pictures, numbers, or number combinations to solve the problem?

- How does the student go about finding the total number of animals? Does she count the total number of pictures in her representation? Does she count counters or fingers to help her find the total? Does she use number combinations that she already knows?

■ How does the student record her answers? Does she use pictures, words, numbers, or a combination of pictures, words, and numbers?

Tasks 4A and 4B

In this unit, children have worked with cube patterns, pattern-block patterns, and clapping patterns. It is expected that they now understand and recognize what a pattern is and can use this knowledge to create, describe, and extend patterns. Additionally, many first graders will be able to identify the unit that repeats in a given pattern. The assessment tasks outlined on Assessment Master 4 give you a chance to see how your students think about patterns.

NOTE

If students are having difficulty continuing the pattern in Problem 4A, you might create a simpler pattern for them to continue, such as the following:

●●■●●■●● ＿ ＿ ＿ ＿

Name _____   Date _____

Assessment Master 4

End-of-Unit Assessment Tasks

4A. Continue this pattern:

●●■▲●●■▲●● __ __ __ __ __ __

4B. Now make a different pattern of your own.

© Scott Foresman, Grade 1   **39**   *Mathematical Thinking at Grade 1*

What to look for ...

■ Can the student continue the pattern? (See margin note.)

■ Can the student create different patterns? How simple or complex are his patterns?

■ Can the student record his own patterns accurately?

YOUR NOTES

_____

_____

_____

_____

Task 5

When first graders begin representing data in their own ways, they often create unique and effective ways of communicating the results of a survey. The main goal of making representations is to show clearly what you found out, so that someone else can understand it. In this unit, students have had many experiences doing different surveys using two categories. This has allowed them to focus on the relationship between the subgroups and the total group. They have learned to make representations of their surveys using Kid Pins, cubes, drawings, or stick-on materials. They have also created representations of their survey data, using three to five categories, to describe how they got to school. It is expected that by the end of this unit, students are able to create their own representations using a few categories.

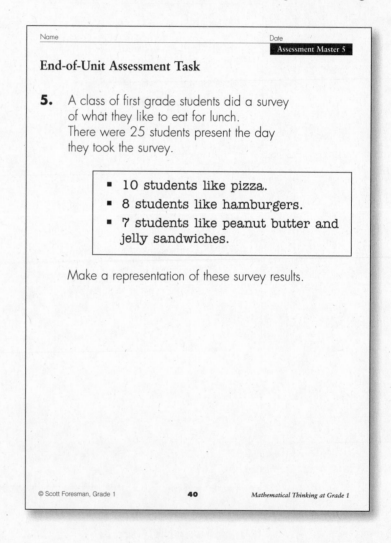

Name _____   Date _____

Assessment Master 5

End-of-Unit Assessment Task

5. A class of first grade students did a survey
of what they like to eat for lunch.
There were 25 students present the day
they took the survey.

> - 10 students like pizza.
> - 8 students like hamburgers.
> - 7 students like peanut butter and jelly sandwiches.

Make a representation of these survey results.

© Scott Foresman, Grade 1   **40**   *Mathematical Thinking at Grade 1*

What to look for ...

- Does the student's representation clearly show the categories and the number of people in each category?

- Does the student make clear visual distinctions between and among categories?

- Does the student's representation make it easy to see how many items are in each category?

- Does the student use pictures, numbers, words, or a combination thereof?

Mathematical Emphases*

1 Uses mathematical materials and tools to solve problems

2 Describes, compares, and finds relationships among geometric shapes

3 Counts and keeps track of a set of up to about 20 objects

4 Orders a set of numbers up to about 20

5 Compares two quantities up to about 20 and can identify which quantity is more and which is less

6 Has a strategy for combining two or more single-digit numbers

7 When combining 2 quantities ...
  a Counts all from 1
  b Counts on from one number
  c Uses numerical reasoning

8 Records solutions to problems using pictures, numbers, and words

9 Finds and records more than one solution to a "How Many of Each?" kind of problem

10 Describes, creates, and extends patterns

11 Classifies patterns as the same or different by identifying the repeating unit within each pattern

12 Collects and records data

13 Categorizes data in ways that communicate clearly to others

14 Creates a representation for a set of data that clearly communicates what a survey is about

15 Makes sense of survey results and presents those results to others

* See **Checklist** on Assessment Master 6.

End-of-Unit Assessment Tasks

Task 1

An important aspect of counting orally is for students to associate number words with the corresponding written numerals. The task presented on Assessment Master 7 is related to counting on the 100 number chart, and at this time students are expected to make transitions between decades correctly.

Name _____ Date _____

Assessment Master 7

End-of-Unit Assessment Task

1. Write numbers in correct sequence on these counting strips:

9
10
11

36
37
38

57
58
59

© Scott Foresman, Grade 1   **42**   *Building Number Sense*

_____

_____

_____

_____

What to look for ...

- Is the student able to write numbers in correct sequence?

- Does the student reverse digits in some numbers? For example, does he write 05 for fifty or 91 for nineteen?

- Is the student able to make transitions between decades? Does he know what comes after 39? After 59?

Task 2

At this point in the Grade 1 sequence, students have had repeated experiences solving "How Many of Each?" problems similar to the one on Assessment Master 8. You might want to compare the students' work on this problem to their work on earlier problems in this unit and in the Mathematical Thinking unit. As you look over their earlier work, look for growth in the students' strategies as well as in the organization of their work. You may want to refer to the "Exploring Multiple Solutions" and "Finding Relationships Among Solutions" Teacher Notes on pages 44–47 of Building Number Sense for additional guidance on examining student work.

Name _____   Date _____

Assessment Master 8

End-of-Unit Assessment Task

2.   Andrew has 13 toys in his toy box.
Some are cars, and some are stuffed animals.
How many of each kind of toy might Andrew have?
Show at least three solutions to this problem
(or show as many ways as you can).

Use pictures, numbers, or words to show your work.

What to look for ...

- Are the student's responses accurate?

- What strategies does the student have for approaching the problem? Does she:
  - ➤ choose to model the problem with cubes or counters?
  - ➤ draw pictures?
  - ➤ mentally know the combinations of 13 and record (some of) those combinations using equations?

- If the student uses counters, does she randomly take counters of two different colors, count them, and then adjust the numbers so that she has thirteen in all? Or does she seem to have a systematic plan for making thirteen by taking away a quantity from one group and then adding on the same quantity to the other group?

- Can the student easily keep track of the total number of cars and animals as well as the individual number of each?

- How does the student record her solutions on paper? Does she draw pictures of cars and animals? Draw squares or other symbols to represent the counters she used to model the problem? Write numbers? Write equations? Write words? Does the student label how many objects are cars and how many are animals?

- Is the student able to think of multiple solutions to the problem? If so, does it appear that her solutions are random? Did the student find pairs of "opposite" solutions (e.g., 1 car and 8 animals; 8 animals and 1 car)? Did the student create ordered lists? Does she recognize and use a numerical pattern? Is she able to find all the different possibilities? Does she have a way of checking whether she has found all the different possibilities?

Task 3

NOTE

Solution strategies might include comparing the total amounts for both pairs of cards or comparing the amounts on individual cards.

Since students have been playing the games "Compare" and "Double Compare," it is expected that they have developed strategies for figuring out which combination of two single-digit numbers is greater or lesser than another combination. If you have the opportunity to ask individual students how they determined their answers, this will give you more information about their evolving strategies for combining and comparing numbers. See the "Double Compare: Strategies for Combining and Comparing" Teacher Note on pages 56–58 of Building Number Sense.

Name _____ Date _____

Assessment Master 9

End-of-Unit Assessment Task

3. Which two numbers have the greater total? Put a circle around them.

6   5   4   9

What to look for ...

- Can the student read and interpret the numerals on the cards, or does he count the objects on the cards to figure out the meanings of the different numbers? Does he use fingers or counters?

- How does the student combine the two quantities (on the two cards in each pair)? Does he count all the objects? Does he count up from one of the numbers to the other number? Does he use his knowledge of number combinations?

YOUR NOTES

- What strategies does the student use for determining which pair's total is greater? Does he know that 13 is greater than 11? Does he use counters to find out or to verify that 13 is greater?

- Does the student reason about which pair's total is greater, without actually finding each total? (I know that 4 + 9 is greater than 6 + 5 because 9 is 3 more than 6, and 5 is only one more than 4.)

Task 4

Students have been solving combining and separating problems in this unit. You might want to compare students' work on the problems presented on Assessment Master 10 to their work on other story problems in this unit. As you look back over their earlier work with story problems, do you see growth in the students' strategies for solving such problems? The "Three Approaches to Story Problems" Teacher Note on pages 133–135 of Building Number Sense *offers suggestions for examining student work.*

Name _____ Date _____

Assessment Master 10

End-of-Unit Assessment Task

4. Solve each problem. Use pictures, numbers, or words to show how you got your answer.

Ayumi and Alex baked some cookies. Ayumi made 7 cookies, and Alex made 8 cookies. How many cookies did they make in all?

There were 13 children on the bus. At the next stop, 5 children got off. How many children were still on the bus?

© Scott Foresman, Grade 1   **45**   *Building Number Sense*

_____

_____

_____

_____

What to look for ...

- Does the student solve the problems correctly?

- How does the student solve each problem? Does she use direct modeling, counting up or counting down, or numerical relationships?

- Is the student using number relationships she knows, such as doubles (e.g., $7 + 7 = 14$ and $14 + 1 = 15$, or $8 + 8 = 16$ and $16 - 1 = 15$, in the first problem on the master)? Or is she breaking apart one number in order to create a multiple of 10 (e.g., $13 - 5$: $5 = 3 + 2$; $13 - 3 = 10$, and $10 - 2 = 8$, in the second problem on the master)?

- How did the student record her work? Can she clearly and accurately record her strategies on paper? Does she use numbers, pictures, words, or equations? Does she use numbers and equations as parts of her recording?

Mathematical Emphases*

1 Accurately counts a set of up to 40 objects

2 Reads, writes, and sequences numbers to 100

3 Associates number words with corresponding written numerals

4 Uses numerals to record how many, for quantities up to 40

5 Finds combinations of numbers up to 15

6 Finds the total of two quantities up to 20

7 Finds the greater of two quantities, up to about 40

8 Knows combinations of 10 (6 + 4, 8 + 2, etc.)

9 Records problem-solving strategies using pictures, numbers, words, and equations

10 Makes, describes, and extends repeating patterns using a variety of materials (e.g., physical actions, objects, drawings, and numbers)

11 Solves story problems involving addition using:
a direct modeling
b counting up/counting down
c numerical relationships

12 Solves story problems involving subtraction using:
a direct modeling
b counting up/counting down
c numerical relationships

13 Finds the total of several small single-digit numbers

14 Finds different combinations for one number (e.g., $12 = 7 + 5$, $8 + 4$, and $3 + 9$)

* See **Checklist** on Assessment Master 11.

End-of-Unit Assessment Tasks

Tasks 1A and 1B

Students have been sorting buttons, lids, and other objects. It is expected that students are able to group sets of objects that are alike in some way. They should be able to choose a clear category and sort the whole set of objects according to this category. (See margin note.) "The Sorting into Groups" Teacher Note on pages 22 and 23 of Survey Questions and Secret Rules offers additional information about how students create and identify categories.

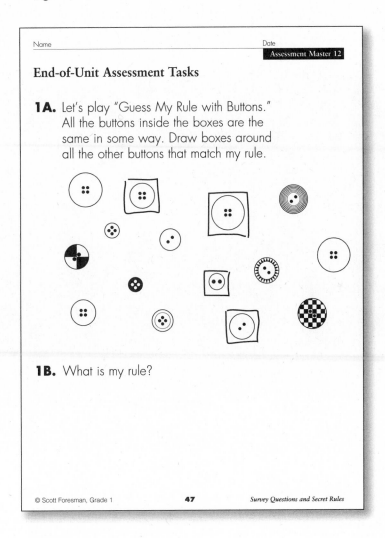

Name _____ Date _____

Assessment Master 12

End-of-Unit Assessment Tasks

1A. Let's play "Guess My Rule with Buttons."
All the buttons inside the boxes are the
same in some way. Draw boxes around
all the other buttons that match my rule.

1B. What is my rule?

© Scott Foresman, Grade 1 47 *Survey Questions and Secret Rules*

What to look for ...

■ Can the student identify the attribute to sort by?

■ Does the student group the buttons in a way that follows the given rule? *(Plain, white buttons)*

Task 2

Students have been sorting many different objects. It is expected that students are able to sort a given collection of objects in more than one way and that they are able to choose a clear category and sort the objects according to their rule.

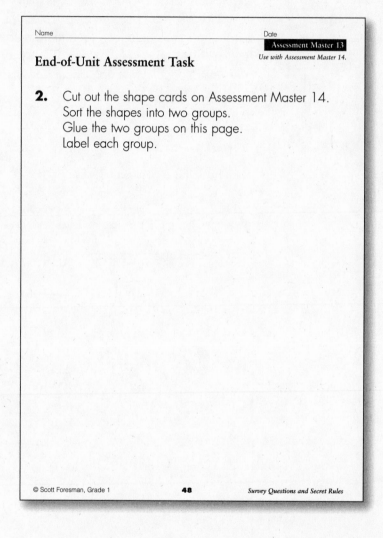

Name _____   Date _____

Assessment Master 13
Use with Assessment Master 14.

End-of-Unit Assessment Task

2. Cut out the shape cards on Assessment Master 14.
Sort the shapes into two groups.
Glue the two groups on this page.
Label each group.

© Scott Foresman, Grade 1   **48**   *Survey Questions and Secret Rules*

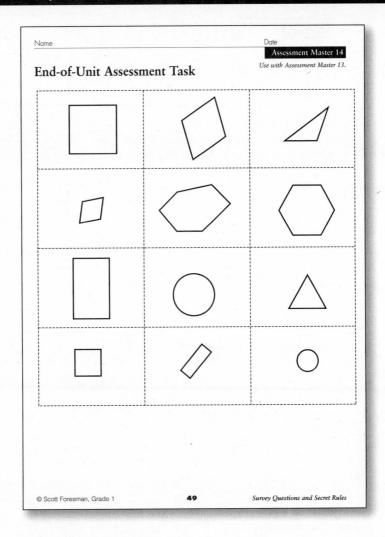

Name _____ Date _____

Assessment Master 14
Use with Assessment Master 13.

End-of-Unit Assessment Task

© Scott Foresman, Grade 1   49   *Survey Questions and Secret Rules*

What to look for ...

- Can the student choose a clear category and sort an entire set of objects according to the rule he has chosen?
- Are the shapes sorted into clear groups, so that you can tell which shapes belong in each group?
- Is the student identifying a single attribute to sort by?
- Is the student able to apply the rule to all the shapes that he is sorting?
- Does the student's written rule (as indicated by his labels) reflect how he chose to group his shapes?

YOUR NOTES

_____

_____

_____

_____

Tasks 3A and 3B

Students have been doing their own surveys in this unit and are familiar with using a class list. Students should recognize that the number of cones and the number of cups add up to the total number of students who voted. See the "Inventing Representations" Teacher Note on pages 40 and 41 of Survey Questions and Secret Rules *for examples of how students organize and represent data.*

Name _____ Date _____

End-of-Unit Assessment Tasks

3A. This survey was done with a class of first graders. Make a representation for the data on a separate sheet.

> **Survey question:**
> Would you rather eat ice cream in a **cone** or in a **cup**?
>
> Alan ▢   Shong ▢   Paul ▼   Pamela ▼
>
> Sara ▼   Masahiro ▢   Joan ▢   Susan ▼
>
> Sean ▢   Linda ▼   Keri ▼   Kindra ▢
>
> Janet ▢   John ▢   Scott ▢   Yuri ▼
>
> Mark ▼   Niki ▢   Allysa ▼

3B. Tell a story about your representation. What did you learn from it?

_____

_____

_____

_____

_____

50   *Survey Questions and Secret Rules*

_____

_____

_____

_____

What to look for ...

- Is the student's representation clear? Does her work clearly communicate what the data set is about?

- Does the student use pictures, words, or numbers in her representation?

- Is the student able to keep track of the number of pieces of data in each group? Does she count accurately? Have all the data been accounted for?

- Are the student's categories visually distinctive? Is it possible to tell from the student's representation that one group (people who prefer cones) is bigger than the other group (people who prefer cups) without counting?

- Does the student's story describe the data both quantitatively and qualitatively?
- As you look over the student's earlier work, do you see growth in her ability to organize and represent data?

Mathematical Emphases*

1 Identifies and describes attributes of various materials

2 Uses one attribute as a basis for sorting and categorizing

3 Looks at a set of grouped objects and identifies the common attribute

4 Is able to make a plan for gathering and recording data

5 Collects and keeps track of data

6 Sorts and categorizes data

7 Invents and constructs representations of data

8 Explains and interprets results of surveys

9 Makes sense of other students' representations

10 Is familiar with calendar features

11 Organizes and orders data about birthdays

12 Creates a timeline representation

13 Can create a representation of data involving several categories

14 Describes data quantitatively and qualitatively

15 Interprets data sets that show values and categories at the same time

16 Compares two data sets

* See **Checklist** on Assessment Master 16.

End-of-Unit Assessment Tasks

Tasks 1A, 1B, and 1C

As students work on the problems outlined on Assessment Masters 17 and 18, it will be important to listen to and observe them as they work to get a sense of how fluent they are becoming in composing and decomposing shapes. Visualizing how a larger unit can be composed of smaller ones, and how shapes can be combined to form larger units, is the basis for understanding many geometric ideas. Many first graders will be able to recognize which shapes will fit different pieces of the outline. They will be able to use equivalents (e.g., 6 triangles for 1 hexagon) to fill the same outline again with a greater or lesser number of pattern blocks. Some first graders, however, may not easily see these relationships and will use a trial-and-error approach.

Name _____   Date _____

Assessment Master 17

End-of-Unit Assessment Tasks   *Use with Assessment Master 18.*

1A. Use pattern blocks to fill in the puzzle on this page.
Then record the shapes you used
by gluing down paper pattern-block shapes.
Record the number of blocks you used.

Number of blocks: _____

© Scott Foresman, Grade 1   **52**   *Quilt Squares and Block Towns*

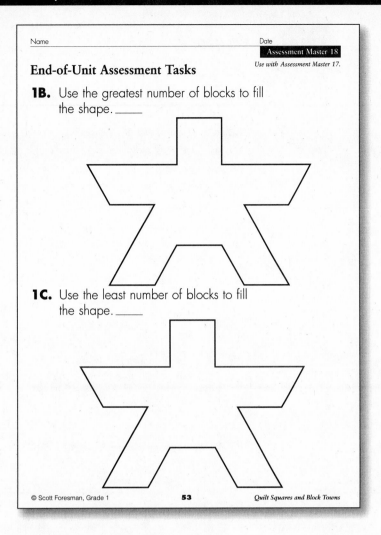

Name _____ Date _____

Assessment Master 18
Use with Assessment Master 17.

End-of-Unit Assessment Tasks

1B. Use the greatest number of blocks to fill
the shape. _____

1C. Use the least number of blocks to fill
the shape. _____

What to look for ...

- Is the student able to fill in the shape outline on Assessment Master 17? Does he easily recognize which pattern block shapes will fill the outline?

- Does the student seem to have a plan for filling in the outline? When he places a block, does he think about what he will need to fill the remaining space?

- Does the student have ways of altering his designs so that he uses more or fewer blocks? Is he able to figure out a way to fill the outline using the greatest and least numbers of blocks?

- Does the student use his knowledge of equivalent shapes to help him with his design (e.g., 2 trapezoids = 1 hexagon; and 3 triangles = 1 trapezoid, or 2 triangles = 1 rhombus)?

- Is the student able to accurately count and record the numbers of blocks that he used?

Task 2

As students work through the problem explained on Assessment Master 19, they should be paying attention to important attributes that shapes have in common, such as which shapes have straight sides, which ones have curves, and which ones have "points" (angles). As they describe these attributes and shapes, students may or may not use correct mathematical vocabulary. As first grade students sort the shapes, many will do so in a way that corresponds to the conventional ways of classifying shapes. But the important thing is that students are carefully observing, describing, and comparing the shapes. See the "Sorting the Shape Cards" Teacher Note on pages 51–53 of Quilt Squares and Block Towns for additional information.

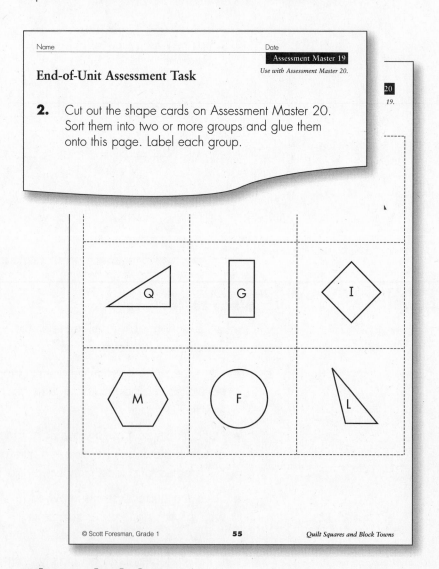

Name _____   Date _____

Assessment Master 19

Use with Assessment Master 20.

End-of-Unit Assessment Task

2. Cut out the shape cards on Assessment Master 20. Sort them into two or more groups and glue them onto this page. Label each group.

© Scott Foresman, Grade 1   **55**   *Quilt Squares and Block Towns*

What to look for ...

- Is the student able to see how the shapes are similar and different?

- Does the student observe and describe the following attributes: number of sides; straight or curved sides; and number of corners?

YOUR NOTES

- Does the student observe and describe differences among shapes that have the same number of sides?

- Is the student able to name her own sorting categories?

- Is the student flexible in thinking about her categories? Can she regroup the shapes if need be?

- Are the shapes placed into accurate categories?

Task 3

You will find a range in your students' abilities to draw and recognize the 3-dimensional shapes of the Geoblocks. Students should recognize that the blocks have different faces that are visible and work to represent these faces in their drawings. Some first graders will show very clearly which blocks are used for their designs and will discuss their relationships to one another. Many first graders will draw the faces of Geoblocks as if they are two-dimensional rather than three-dimensional shapes. They may be able to show how the blocks are arranged, but not capture the three-dimensionality of the Geoblocks. Some first graders may have difficulty drawing the shapes and showing how the blocks go together. Assessment should focus not on students' drawing abilities, but on their abilities to observe, describe, and compare 3-D shapes. The "Students Draw in 3-D" Teacher Note on pages 106 and 107 of Quilt Squares and Block Towns *offers further information about how students approach the drawing of 3-dimensional constructions.*

Name _____ Date _____

Assessment Master 21

End-of-Unit Assessment Task

3. Build something using 3 or 4 Geoblocks.
Draw what you made.

What to look for ...

- Does the student demonstrate in his drawings that each Geoblock has more than one face? Does the student notice that some Geoblocks have only rectangular faces while others have triangular faces? Does he distinguish between these faces in his drawings?

- Is the student able to effectively show how the blocks are arranged? Is he able to show how the blocks go together?

- Does the student notice that some Geoblocks are somewhat alike in shape and can be distinguished by comparing dimensions (size and thickness)? In the student's drawing, for example, does he distinguish between a large cube and a small cube, or represent one rectangular prism as much thinner than another?

- Is the student aware of how the faces are connected to make a solid object?

Task 4

In this unit, students have had experiences constructing their own boxes. It is expected that they will be able to visualize what will make a box. Some students are able to come up with a clear plan for constructing their boxes, laying the pieces out ahead of time to see how they can "fold up" their boxes. These students have a clear understanding that the opposite sides of a prism are the same size. There may be some students who will begin to use this approach, but have difficulty finding all the pieces necessary to make a box. Other students will be able to fit together portions of the box, but have no pieces that will fit the last two sides. Still others will not be able to coordinate the pieces to make a rectangular prism. See the "Students Create Their Own Boxes" Teacher Note on pages 96 and 97 of Quilt Squares and Block Towns for examples and descriptions of how first graders approach this activity.

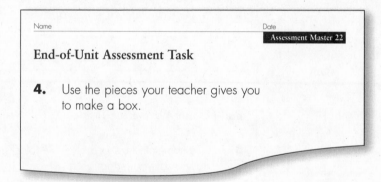

Name _____   Date _____

Assessment Master 22

End-of-Unit Assessment Task

4.   Use the pieces your teacher gives you
to make a box.

What to look for ...

- Does the student make use of the information that the opposite faces of a rectangular prism are the same size and length? Does she choose adjacent faces with edges of equal lengths?

- Does the student have ways to approach the problem? Does she lay out the pieces on the table and see how she can "fold" them up into a box?

- When a student begins a box that doesn't work, does she seem stuck, or does she then try a different combination of pieces? Does the "stuck" student learn from what doesn't work how to make better choices—or does she start again, choosing pieces randomly?

- When the student runs into a problem, can she figure out a way to overcome the problem and finish her box, even if the result is not a conventional box?

NOTE

The pieces for this assessment activity must be cut ahead of time and placed in envelopes. The preparation is similar to the Making Boxes activity in Sessions 8–10 of Investigation 2. We recommend using index cards for this activity as well. There should be one envelope per student. Each envelope should contain the following:

two 5 × 8 pieces;

four 2 × 8 pieces;

two 2 × 5 pieces;

two 2 × 3 pieces;

two 3 × 8 pieces.

Students use the given pieces to construct their own boxes.

YOUR NOTES

Mathematical Emphases*

2-Dimensional Shapes

1 Observes, describes, and compares shapes

2 Uses mathematical vocabulary to describe and name shapes

3 Describes characteristics of triangles

4 Groups shapes according to common characteristics

5 Composes and decomposes shapes

6 Notices relationships among shapes

7 Uses rotation and reflection to arrange shapes

8 Fills a given region with shapes

9 Visualizes and represents shapes

10 Is familiar with a variety of squares, rectangles, and triangles

3-Dimensional Shapes

11 Constructs, observes, describes, and compares shapes and objects

12 Uses mathematical vocabulary to describe shapes

13 Constructs 3-D shapes from 2-D shapes

14 Visualizes and describes rectangular prisms

15 Compares and describes objects by size, shape, and orientation

16 Puts 3-D shapes together to make other shapes

17 Creates and uses 2-D representations of 3-D shapes and objects

18 Notices shapes in the environment

Paths and Patterns

19 Visualizes, describes, and compares the path between two locations in space and on a grid

20 Estimates distances

21 Visualizes and describes directions of turns

22 Follows, gives, and records directions for how to move in space and on a path

23 Builds a pattern by repeating a unit square

24 Sees how changing the unit affects the whole pattern

* See **Checklist** on Assessment Master 23.

End-of-Unit Assessment Tasks

Task 1

Students differ widely in how readily they make sense of and solve the type of problem presented on Assessment Master 24. Some may make different combinations of red and blue counters until they find one that works; others may use their knowledge of number combinations; and still others may find new solutions by changing something about the solution they already have. Some students may record only one or two responses, while others will be able to find all the possible solutions. Students will also vary in their ability to organize their responses. You might want to compare the students' work on this problem to their work on other crayon puzzles in this unit. As you look over the earlier work, do you see growth in the students' strategies for solving problems and in their ability to organize their work?

Name _____   Date _____

Assessment Master 24

End-of-Unit Assessment Task

1. I have 13 crayons.
   Some are blue and some are red.
   I have more blue crayons than red crayons.
   How many of each color might I have?

What to look for ...

- Is the student's work accurate?

- Does the student understand what he is to find? Can he keep all the clues in mind?

- What strategies does the student have for solving the problem? Does he:

  ➤ generate combinations randomly until he finds one that matches all the clues?

  ➤ use his knowledge of number combinations?

  ➤ build on his experience with previous crayon puzzles?

  ➤ assemble a set of blue and red counters that matches the total, and then adjust the counters (the subtotals) until he finds a solution with the appropriate amounts of each color?

- Can the student find more than one solution?

- Does the student treat each new solution as a separate problem, or does he find new solutions by changing something about the solutions he already has? Does the student use a pattern to find new solutions?

- Can the student find all the possible solutions to the problem?

- Is the student able to organize his work?

Task 2

When working on this problem, some students may use strategies that include grouping and counting by 2's, using number combinations, and using doubles. However, some students may still feel most comfortable directly modeling the problem and counting by ones.

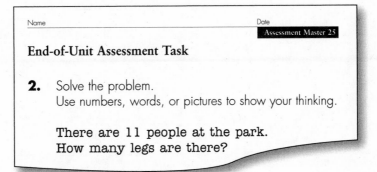

Name _____ Date _____

Assessment Master 25

End-of-Unit Assessment Task

2. Solve the problem.
Use numbers, words, or pictures to show your thinking.

There are 11 people at the park.
How many legs are there?

YOUR NOTES

What to look for ...

- Is the student's answer correct?

- Is the student able to make sense of the problem?

- What strategy did the student use to solve the problem? Did she:

  ➤ model the problem with pictures or objects?

  ➤ use strategies based on counting by 1's? 2's? 4's?

  ➤ use strategies based on number combinations or doubles?

  ➤ break the problem into smaller parts and then combine the parts?

Task 3

Since the beginning of the year, students have had lots of experience with the type of problem presented on Assessment Master 26. Some may still solve story problems using direct modeling, showing the actions in the problem step by step in order to solve it. However, many students will have developed a better understanding of numbers and the relationships among them, and they will use strategies based on counting and numerical reasoning. Students begin counting on or counting down, instead of beginning from 1. Some use what they know about number combinations to find solutions, and some may use numerical reasoning by breaking numbers apart into useful chunks, manipulating those chunks, and then putting them back together. You might want to compare your students' work on this problem to their other work on other story problems in this unit and throughout the year. As you look over their earlier work, do you see growth in their strategies for solving problems and in their ability to organize their work?

Name _____   Date _____

Assessment Master 26

End-of-Unit Assessment Task

3. Solve the problem.
   Use numbers, words, or pictures to show your thinking.

   There were 13 ducks and 9 geese at the
   park. How many animals were there?

What to look for ...

- Is the student's answer correct?

- Is the student able to identify the problem as a *combining* situation? Did he model it with pictures or objects, or did he work mentally?

- What strategies does the student use? Does he:

  ➤ count out a set of 13 and a set of 9 and then count all the objects?

  ➤ count on from one number, keeping track of the number added as he counts?

  ➤ use his knowledge of number combinations and relationships (e.g., 13 + 10 = 23, so 13 + 9 = 22)?

- Can the student clearly and accurately record his thinking on paper?

- Is the student using numbers and equations as parts of his explanation?

Task 4

You might want to compare your students' work on the problem presented on Assessment Master 27 to their work on other story problems in this unit and throughout the year. As you look over their earlier work, do you see growth in their strategies for solving problems and in their ability to organize their work?

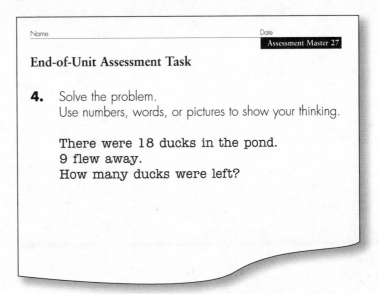

Name _____   Date _____

Assessment Master 27

End-of-Unit Assessment Task

4.   Solve the problem.
Use numbers, words, or pictures to show your thinking.

There were 18 ducks in the pond.
9 flew away.
How many ducks were left?

YOUR NOTES

What to look for ...

- Is the student's answer correct?

- Does the student understand that quantities are being separated, and not combined?

- How does the student approach the problem? Does she model it with pictures or objects, or does she work mentally?

- What strategies is the student using? Does she:

  ➤ count out a set of 18 and then remove 9?

  ➤ count back from 18 or count up from 9?

  ➤ use her knowledge of number combinations and relationships (e.g., $18 - 10 = 8$, so $18 - 9 = 9$ or $9 + 9 = 18$)?

- Can the student clearly and accurately record her thinking on paper?

- Does the student use numbers and equations as parts of her explanation?

Mathematical Emphases*

1 Reads, writes, and sequences numbers to 100

2 Finds the total of two or more single-digit numbers

3 Knows combinations of 10
(e.g., 6 + 4, 8 + 2, and 7 + 3)

4 Is increasing his or her familiarity with single-digit addition pairs

5 Finds combinations of numbers up to 20

6 Understands more, less, and equal amounts

7 Uses equations to describe arrangements of objects or pictures in groups

8 Analyzes visual images of quantities

9 Records solutions with pictures, numbers, words, and equations

10 Finds more than one solution to a problem that has multiple solutions

11 Is developing a strategy for organizing sets of objects so that they are easier to count and combine

12 Is developing meaning for counting by 2's

13 Is becoming familiar with coin names, values, and equivalencies

14 Identifies, describes, creates, and extends repeating patterns

15 Represents a pattern in different ways (i.e., with physical actions, concrete materials, drawings, and/or numbers)

16 Identifies some patterns in the number sequence and on the 100 chart

17 Uses the calculator as a mathematical tool

18 Solves story problems involving addition using:
a direct modeling
b counting up/counting down
c numerical relationships

19 Solves story problems involving subtraction using:
a direct modeling
b counting up/counting down
c numerical relationships

20 Can visualize story problems that involve combining with unknown change

21 Creates story problems to match addition expressions

* See **Checklist** on Assessment Master 28.

End-of-Unit Assessment Tasks

Tasks 1A, 1B, and 1C

In this unit, students have had numerous opportunities to develop their perception of what weight is so that they can distinguish it from other types of measurement. It is expected that they can use balance scales accurately to determine whether objects weigh more than, less than, or the same as a particular object they have chosen. The focus of teacher assessment should be the process of weighing, rather than the numerical result of the process.

Name _____ Date _____

Assessment Master 29

End-of-Unit Assessment Tasks

Put an object on one side of the balance scale.

1A. Put something **heavier** on the other side.
First object: _____
Heavier object: _____
Draw a picture to show how the balance scale looks.

1B. Put something **lighter** on the other side.
First object: _____ Lighter object: _____
Draw a picture to show how the balance scale looks.

1C. Put something that weighs **the same** on the other side.
First object: _____
Equal-weight object: _____
Draw a picture to show how the balance scale looks.

© Scott Foresman, Grade 1   64   *Bigger, Taller, Heavier, Smaller*

What to look for ...

- Are the student's answers accurate?
- Does the student notice that it matters where in the pan the objects are placed in order to balance them?
- Is the student aware that the balance arm needs to stop swinging before he decides whether an object is heavier or lighter than another object?

- Is the student comfortable with the idea that the "down" position on the balance scale indicates a heavier object than the "up" position indicates? Does the student know what it means when the balance scales are even?

- How does the student record the results of balancing? Are his representations clear to others?

Tasks 2A and 2B

Students have had opportunities to fill a 2-dimensional shape outline with pattern blocks. They are expected to be able to determine which sets of given blocks do or do not fit a particular outline. They should also be able to create their own set of blocks to fit the shape outline exactly.

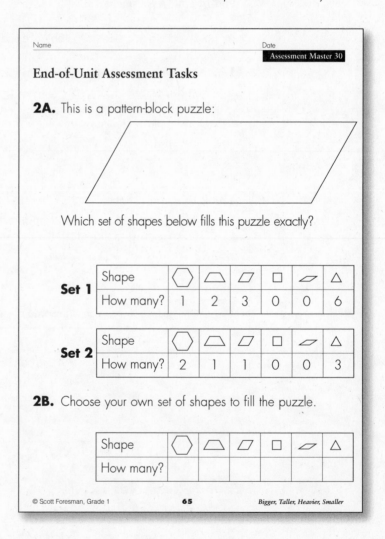

What to look for ...

- How does the student fill the outlined shape at the top of the master? Is she fitting blocks inside the shape exactly? Is she able to judge when blocks are hanging over the edge or when the space is not filled?

- Is the student able to find strategies for fitting shapes into the interior of the outline when it is not obvious which shapes will work?

- When one arrangement of blocks doesn't work, does the student assume the set doesn't fit, or does she experiment with different arrangements? Does she slide the blocks away after filling a shape, and then begin randomly again, or does she try to make revisions to the design she already has in place?

- Does the student use relationships among shapes to reason why one set fills an area and why another set is too big or too small for the area?

- When the student makes her own arrangements of shapes to fit the outline, does she use information about the previous sets to help her create her design? Does she use numerical reasoning to make the same shape in different ways (e.g., a hexagon can be replaced by 6 triangles)?

- Is the student able to accurately fill the outline with her own set of pattern blocks?

Task 3

NOTE

For this assessment activity, students can use the materials assembled for Investigation 2. They should compare the capacities of two empty containers that vary in shape, but that are not drastically different in capacity. Students can choose from among a variety of measurement tools to fill the containers, such as a measuring cup, a spoon, and a small plastic or paper cup. They can also measure with water or with something dry, such as sand, rice, cat litter, or beans.

Students have been measuring and comparing the capacities of different containers using a variety of nonstandard units. It is expected that students should be able to compare the containers directly, or count in nonstandard units the amount needed to fill each one. Students should be learning that an established, consistent unit of measurement will enable them to compare and describe different capacities more precisely and clearly. The focus of teacher assessment should be the process of measuring, rather than the numerical result of the process.

Name _____ Date _____

Assessment Master 31

End-of-Unit Assessment Task

3. Choose 2 containers. Choose a measuring tool. Use the tool to fill each container with sand.

- How much sand does each container hold?
- Which container holds more sand?

Write and draw about how you measured. Tell what you found out.

_____

_____

_____

What to look for ...

- Are the student's answers accurate?

- Is the student able to find a fixed technique for measuring? Is he consistently using full units of measurement?

- Does the student keep track of his measurements in a clear way that makes sense to him? Is he able to count accurately? Does he double-check his counts?

- How does the student compare the capacities of the two containers? Is he measuring each individually and comparing the number of units, or is he pouring the contents of one container into the other?

- How does the student decide when the containers are full?

- How does the student handle the situation if he needs only a part of a measurement unit to finish filling the container? What does he record?

- Can the student determine which container holds more? Can the student make a reasonable statement telling _why_ one container holds more sand than the other container holds?

NOTE

Students can use the materials assembled for Investigation 3 for this activity.

Task 4

Through their experiences with comparing, measuring, and ordering lengths, students should be developing an understanding of what length is and how it can be described. They are expected to be able to align the ends of two objects in order to compare or measure them. Students should begin to be able to iterate (repeat) a nonstandard unit of measurement along a length without gaps or overlaps. Students should also be learning that an established, consistent unit of measurement will enable them to compare and describe different lengths more precisely and clearly. The focus of teacher assessment should be the process of measuring, rather than the numerical result of the process.

Name   Date

Assessment Master 32

End-of-Unit Assessment Task

4. Choose 5 boxes from the "Mystery Box" collection.
Measure the longest side of each box with cubes.
Put the boxes in order from shortest to longest.
Show how you organized your boxes. Include
how many cubes long each box is.

© Scott Foresman, Grade 1   **67**   *Bigger, Taller, Heavier, Smaller*

What to look for ...

- Is the student's work accurate?

- Does the student select the longest dimension to measure?

- Does the student align the beginning or end of the cube tower with the beginning or end of the box she is measuring, or does she forget to align these two things?

- Is the student counting and keeping track of the number of cubes correctly?

- What does the student decide to do when the length is not close to a number of whole cubes? Does she find a way to describe this amount of cubes?

- What strategies does the student use to order the boxes? Does the student use numbers to help her? Does she compare cube towers or the boxes themselves?

- Is the student's representation organized? Does she clearly show all of her boxes ordered by length? Is her representation easy to read? Does she use the layout of her drawings on the page, or the sizes of the drawn boxes, numbers, or words to show the order?

Mathematical Emphases*

Understands what weight is

1 Demonstrates a sense of *heavy* and *light* by feel

2 Uses language to describe and compare weights

3 Uses a balance to weigh objects and compares the weights of different objects using a balance

4 Represents the results of weight comparisons

Understands what capacity is

5 Uses language to describe and compare capacities

6 Compares and measures capacities using nonstandard units

7 Collects, keeps track of, and interprets data

8 Estimates the number of units needed to fill a container

9 Counts and keeps track of quantities up to 50

10 Relates size and shape to capacity

11 Compares the capacities of two containers by filling them with continuous substances (such as water) or discrete objects (such as cubes)

12 Compares the capacities of more than two containers

13 Fills a given area with shapes

Understands what length is

14 Uses language to describe and compare lengths

15 Uses direct comparison of lengths

16 Measures and compares lengths using nonstandard units

17 Orders lengths

18 Represents measurements with numbers, concrete materials, and/or pictures in a clear, ordered way

19 Develops, describes, and justifies techniques for filling containers, comparing capacities, and measuring lengths

20 Describes measurements that can't be measured in whole, exact units

* See **Checklist** on Assessment Master 34.

Assessment Masters

End-of-Unit Assessment Tasks

1A. Draw 14 circles in the space below.

1B. How many more circles do you need to draw to have 20 in all? Add these to the circles you have drawn.

End-of-Unit Assessment Task

2. You have 9 shapes. Some of them are circles, and some of them are squares. How many of each shape could you have? Find as many different solutions as you can.

End-of-Unit Assessment Task

3. At the zoo, the children saw many animals.
There were 2 monkeys on a tree,
4 tigers on the ground,
3 lions in a den,
and 2 zebras inside a fenced area.
How many animals did the children see?

End-of-Unit Assessment Tasks

4A. Continue this pattern:

4B. Now make a different pattern of your own.

End-of-Unit Assessment Task

5. A class of first grade students did a survey of what they like to eat for lunch. There were 25 students present the day they took the survey.

> - 10 students like pizza.
> - 8 students like hamburgers.
> - 7 students like peanut butter and jelly sandwiches.

Make a representation of these survey results.

✓ Checklist of Mathematical Emphases for *Mathematical Thinking at Grade 1*	Student's Name	Student's Name	Student's Name
1 Uses mathematical materials and tools to solve problems			
2 Describes, compares, and finds relationships among geometric shapes			
3 Counts and keeps track of a set of up to about 20 objects			
4 Orders a set of numbers up to about 20			
5 Compares two quantities up to about 20 and can identify which quantity is more and which is less			
6 Has a strategy for combining two or more single-digit numbers			
7 When combining 2 quantities … **a** Counts all from 1 **b** Counts on from one number **c** Uses numerical reasoning			
8 Records solutions to problems using pictures, numbers, and words			
9 Finds and records more than one solution to a "How Many of Each?" kind of problem			
10 Describes, creates, and extends patterns			
11 Classifies patterns as the same or different by identifying the repeating unit within each pattern			
12 Collects and records data			
13 Categorizes data in ways that communicate clearly to others			
14 Creates a representation for a set of data that clearly communicates what a survey is about			
15 Makes sense of survey results and presents those results to others			

End-of-Unit Assessment Task

1.   Write numbers in correct sequence
on these counting strips:

9		36		57
10		37		58
11		38		59

End-of-Unit Assessment Task

2. Andrew has 13 toys in his toy box.
Some are cars, and some are stuffed animals.
How many of each kind of toy might Andrew have?

Show at least three solutions to this problem
(or show as many ways as you can).

Use pictures, numbers, or words to show your work.

End-of-Unit Assessment Task

3. Which two numbers have the greater total?
Put a circle around them.

End-of-Unit Assessment Task

4.   Solve each problem. Use pictures, numbers,
or words to show how you got your answer.

Ayumi and Alex baked some cookies.
Ayumi made 7 cookies, and Alex made
8 cookies. How many cookies did
they make in all?

_____

There were 13 children on the bus.
At the next stop, 5 children got off.
How many children were still on the bus?

✓ Checklist of Mathematical Emphases for *Building Number Sense*

	Student's Name	Student's Name	Student's Name
1 Accurately counts a set of up to 40 objects			
2 Reads, writes, and sequences numbers to 100			
3 Associates number words with corresponding written numerals			
4 Uses numerals to record how many, for quantities up to 40			
5 Finds combinations of numbers up to 15			
6 Finds the total of two quantities up to 20			
7 Finds the greater of two quantities, up to about 40			
8 Knows combinations of 10 (6 + 4, 8 + 2, etc.)			
9 Records problem-solving strategies using pictures, numbers, words, and equations			
10 Makes, describes, and extends repeating patterns using a variety of materials (e.g., physical actions, objects, drawings, and numbers)			
11 Solves story problems involving addition using: **a** direct modeling **b** counting up/counting down **c** numerical relationships			
12 Solves story problems involving subtraction using: **a** direct modeling **b** counting up/counting down **c** numerical relationships			
13 Finds the total of several small single-digit numbers			
14 Finds different combinations for one number. (e.g., 12 = 7 + 5, 8 + 4, and 3 + 9)			

End-of-Unit Assessment Tasks

1A. Let's play "Guess My Rule with Buttons."
All the buttons inside the boxes are the
same in some way. Draw boxes around
all the other buttons that match my rule.

1B. What is my rule?

End-of-Unit Assessment Task

2. Cut out the shape cards on Assessment Master 14.
Sort the shapes into two groups.
Glue the two groups on this page.
Label each group.

End-of-Unit Assessment Task

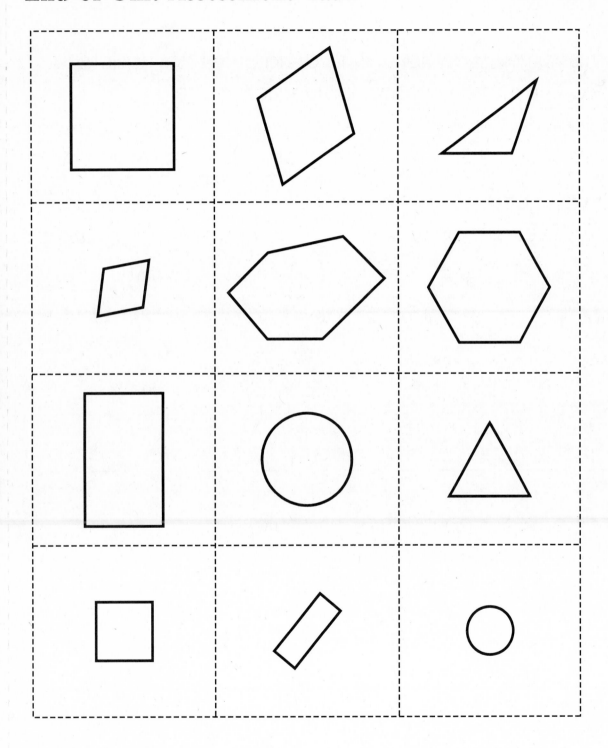

End-of-Unit Assessment Tasks

3A. This survey was done with a class of first graders. Make a representation for the data on a separate sheet.

Survey question:
Would you rather eat ice cream in a **cone** or in a **cup**?

Alan 🥤	Shong 🥤	Paul 🍦	Pamela 🍦
Sara 🍦	Masahiro 🥤	Joan 🥤	Susan 🍦
Sean 🍦	Linda 🍦	Keri 🍦	Kindra 🥤
Janet 🥤	John 🥤	Scott 🥤	Yuri 🍦
Mark 🍦	Niki 🥤	Allysa 🍦	

3B. Tell a story about your representation. What did you learn from it?

_____

_____

_____

_____

_____

Checklist of Mathematical Emphases for Survey Questions and Secret Rules ✓	Student's Name	Student's Name	Student's Name
1 Identifies and describes attributes of various materials			
2 Uses one attribute as a basis for sorting and categorizing			
3 Looks at a set of grouped objects and identifies the common attribute			
4 Is able to make a plan for gathering and recording data			
5 Collects and keeps track of data			
6 Sorts and categorizes data			
7 Invents and constructs representations of data			
8 Explains and interprets results of surveys			
9 Makes sense of other students' representations			
10 Is familiar with calendar features			
11 Organizes and orders data about birthdays			
12 Creates a timeline representation			
13 Can create a representation of data involving several categories			
14 Describes data quantitatively and qualitatively			
15 Interprets data sets that show values and categories at the same time			
16 Compares two data sets			

End-of-Unit Assessment Tasks

1A. Use pattern blocks to fill in the puzzle on this page.
Then record the shapes you used
by gluing down paper pattern-block shapes.
Record the number of blocks you used.

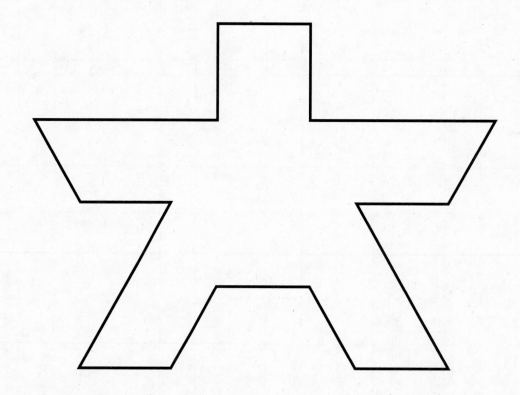

Number of blocks: _____

End-of-Unit Assessment Tasks

1B. Use the greatest number of blocks to fill the shape. _____

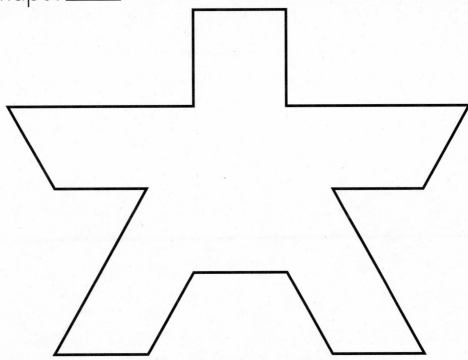

1C. Use the least number of blocks to fill the shape. _____

End-of-Unit Assessment Task

2. Cut out the shape cards on Assessment Master 20. Sort them into two or more groups and glue them onto this page. Label each group.

End-of-Unit Assessment Task

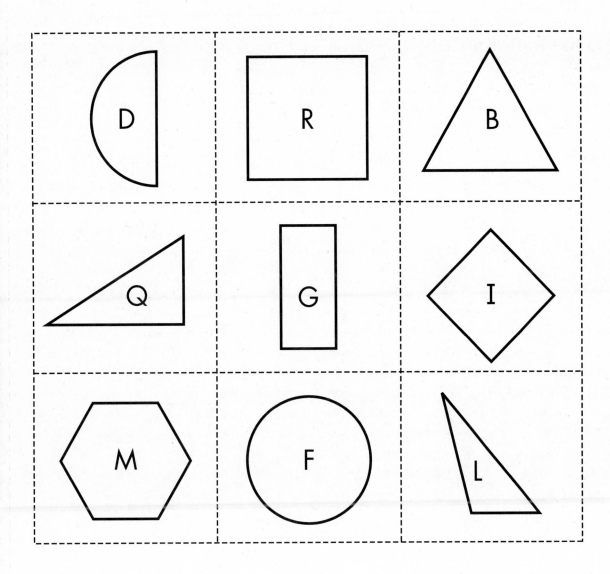

End-of-Unit Assessment Task

3. Build something using 3 or 4 Geoblocks.
Draw what you made.

End-of-Unit Assessment Task

4. Use the pieces your teacher gives you
to make a box.

✓ Checklist of Mathematical Emphases for Quilt Squares and Block Towns

	Student's Name	Student's Name	Student's Name

2-Dimensional Shapes
1 Observes, describes, and compares shapes
2 Uses mathematical vocabulary to describe and name shapes
3 Describes characteristics of triangles
4 Groups shapes according to common characteristics
5 Composes and decomposes shapes
6 Notices relationships among shapes
7 Uses rotation and reflection to arrange shapes
8 Fills a given region with shapes
9 Visualizes and represents shapes
10 Is familiar with a variety of squares, rectangles, and triangles

3-Dimensional Shapes
11 Constructs, observes, describes, and compares shapes and objects
12 Uses mathematical vocabulary to describe shapes
13 Constructs 3-D shapes from 2-D shapes
14 Visualizes and describes rectangular prisms
15 Compares and describes objects by size, shape, and orientation
16 Puts 3-D shapes together to make other shapes
17 Creates and uses 2-D representations of 3-D shapes and objects

Paths and Patterns
18 Notices shapes in the environment
19 Visualizes, describes, and compares the path between two locations in space and on a grid
20 Estimates distances
21 Visualizes and describes directions of turns
22 Follows, gives, and records directions for how to move in space and on a path
23 Builds a pattern by repeating a unit square
24 Sees how changing the unit affects the whole pattern

End-of-Unit Assessment Task

1. I have 13 crayons.
Some are blue and some are red.
I have more blue crayons than red crayons.
How many of each color might I have?

End-of-Unit Assessment Task

2. Solve the problem.
Use numbers, words, or pictures to show your thinking.

There are 11 people at the park.
How many legs are there?

End-of-Unit Assessment Task

3. Solve the problem.
Use numbers, words, or pictures to show your thinking.

There were 13 ducks and 9 geese at the park. How many animals were there?

End-of-Unit Assessment Task

4. Solve the problem.
Use numbers, words, or pictures to show your thinking.

There were 18 ducks in the pond.
9 flew away.
How many ducks were left?

Checklist of Mathematical Emphases for Number Games and Story Problems

	Student's Name	Student's Name	Student's Name	Student's Name
1 Reads, writes, and sequences numbers to 100				
2 Finds the total of two or more single-digit numbers				
3 Knows combinations of 10 (e.g., 6 + 4, 8 + 2, and 7 + 3)				
4 Is increasing his or her familiarity with single-digit addition pairs				
5 Finds combinations of numbers up to 20				
6 Understands more, less, and equal amounts				
7 Uses equations to describe arrangements of objects or pictures in groups				
8 Analyzes visual images of quantities				
9 Records solutions with pictures, numbers, words, and equations				
10 Finds more than one solution to a problem that has multiple solutions				
11 Is developing a strategy for organizing sets of objects so that they are easier to count and combine				
12 Is developing meaning for counting by 2's				
13 Is becoming familiar with coin names, values, and equivalencies				
14 Identifies, describes, creates, and extends repeating patterns				
15 Represents a pattern in different ways (i.e., with physical actions, concrete materials, drawings, and/or numbers)				
16 Identifies some patterns in the number sequence and on the 100 chart				
17 Uses the calculator as a mathematical tool				
18 Solves story problems involving addition using:				
a direct modeling				
b counting up/counting down				
c numerical relationships				
19 Solves story problems involving subtraction using:				
a direct modeling				
b counting up/counting down				
c numerical relationships				
20 Can visualize story problems that involve combining with unknown change				
21 Creates story problems to match addition expressions				

End-of-Unit Assessment Tasks

Put an object on one side of the balance scale.

1A. Put something **heavier** on the other side.

First object:_____

Heavier object:_____

Draw a picture to show how thebalance scale looks.

1B. Put something **lighter** on the other side.

First object:_____   Lighter object:_____

Draw a picture to show how the balance scale looks.

1C. Put something that weighs **the same** on the other side.

First object:_____

Equal-weight object:_____

Draw a picture to show how the balance scale looks.

End-of-Unit Assessment Tasks

2A. This is a pattern-block puzzle:

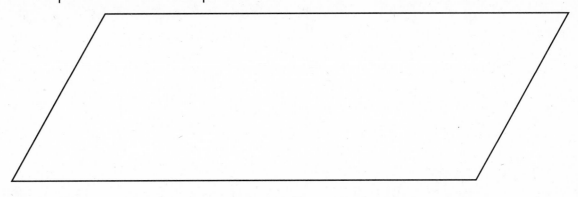

Which set of shapes below fills this puzzle exactly?

Set 1

Shape	⬡	⬯	▱	▢	▱	△
How many?	1	2	3	0	0	6

Set 2

Shape	⬡	⬯	▱	▢	▱	△
How many?	2	1	1	0	0	3

2B. Choose your own set of shapes to fill the puzzle.
Record how many of each shape you used.

Shape	⬡	⬯	▱	▢	▱	△
How many?						

Bigger, Taller, Heavier, Smaller

End-of-Unit Assessment Task

3.   Choose 2 containers. Choose a measuring tool.
Use the tool to fill each container with sand.

- How much sand does each container hold?
- Which container holds more sand?

Write and draw about how you measured.
Tell what you found out.

End-of-Unit Assessment Task

4.   Choose 5 boxes from the "Mystery Box" collection.
Measure the longest side of each box with cubes.
Put the boxes in order from shortest to longest.
Show how you organized your boxes. Include
how many cubes long each box is.

✓ Checklist of Mathematical Emphases for *Bigger, Taller, Heavier, Smaller*

	Student's Name	Student's Name	Student's Name
Understands what weight is			
1 Demonstrates a sense of *heavy* and *light* by feel			
2 Uses language to describe and compare weights			
3 Uses a balance to weigh objects and compares the weights of different objects using a balance			
4 Represents the results of weight comparisons			
Understands what capacity is			
5 Uses language to describe and compare capacities			
6 Compares and measures capacities using nonstandard units			
7 Collects, keeps track of, and interprets data			
8 Estimates the number of units needed to fill a container			
9 Counts and keeps track of quantities up to 50			
10 Relates size and shape to capacity			
11 Compares the capacities of two containers by filling them with continuous substances (such as water) or discrete objects (such as cubes)			
12 Compares the capacities of more than two containers			
13 Fills a given area with shapes			
Understands what length is			
14 Uses language to describe and compare lengths			
15 Uses direct comparison of lengths			
16 Measures and compares lengths using nonstandard units			
17 Orders lengths			
18 Represents measurements with numbers, concrete materials, and/or pictures in a clear, ordered way			
19 Develops, describes, and justifies techniques for filling containers, comparing capacities, and measuring lengths			
20 Describes measurements that can't be measured in whole, exact units			

Multicopias de evaluación

Tareas de evaluación al final de la unidad

1A. Dibuja 14 círculos en el espacio de abajo.

1B. ¿Cuántos círculos más necesitas dibujar para tener 20 en total? Suma éstos a los círculos que has dibujado.

Tarea de evaluación al final de la unidad

2. Tienes 9 figuras.
Algunas de ellas son círculos
y otras son cuadrados.
¿Cuántas figuras de cada tipo
podrías tener?
Halla todas las soluciones que puedas.

Tarea de evaluación al final de la unidad

3. En el zoológico los niños vieron
muchos animales.
En un árbol había 2 monos,
en la tierra había 4 tigres,
en la cueva había 3 leones
y en un área cercada había 2 cebras.
¿Cuántos animales vieron los niños?

Tareas de evaluación al final de la unidad

4A. Continúa este patrón:

4B. Ahora crea tu propio patrón.

Tarea de evaluación al final de la unidad

5. Una clase de estudiantes de primer grado hizo una encuesta de lo que les gusta comer en el almuerzo. El día que hicieron la encuesta había 25 estudiantes presentes.

> - A 10 estudiantes les gusta la pizza.
> - A 8 estudiantes les gustan las hamburguesas.
> - A 7 estudiantes les gustan los sándwiches de mantequilla de maní y jalea.

Haz una representación de los datos de esta encuesta.

✓ Lista de énfasis matemáticos para *Razonamiento matemático para el primer grado*

	Nombre del estudiante	Nombre del estudiante	Nombre del estudiante
1 Usa materiales y herramientas matemáticas para resolver problemas.			
2 Describe, compara y halla relaciones entre figuras geométricas.			
3 Enumera y lleva la cuenta de conjuntos de hasta 20 objetos.			
4 Ordena un conjunto de números hasta 20 aproximadamente.			
5 Compara dos cantidades hasta 20 aproximadamente y puede identificar qué cantidad indica más y cuál indica menos.			
6 Tiene una estrategia para combinar dos o más números de un dígito.			
7 Cuando combina dos cantidades.... **a** cuenta todo a partir de 1. **b** cuenta hacia adelante desde un número determinado. **c** usa razonamiento numérico.			
8 Anota las soluciones de los problemas usando dibujos, números y palabras.			
9 Halla y anota más de una solución para los problemas del tipo "¿Cuántos hay de cada uno?"			
10 Describe, crea y extiende patrones.			
11 Clasifica patrones en iguales o diferentes, identificando la unidad de repetición en cada uno.			
12 Reúne y anota datos.			
13 Categoriza datos de manera que se comuniquen claramente a los demás.			
14 Crea una representación de un conjunto de datos que comunica claramente de qué se trata una encuesta.			
15 Entiende los resultados de las encuestas y los presenta ante los demás.			

Tarea de evaluación al final de la unidad

1. En estas tiras para contar,
escribe números en la secuencia correcta:

9		36		57
10		37		58
11		38		59

Tarea de evaluación al final de la unidad

2. Andrés tiene 13 juguetes en su caja.
Algunos de ellos son autos y otros
son animales.
¿Cuántos juguetes de cada tipo podría
tener Andrés?

Muestra por lo menos tres soluciones
para este problema (o muestra todas
las maneras que puedas).

Usa dibujos, números o palabras para
mostrar tu trabajo.

Tarea de evaluación al final de la unidad

3. ¿Qué par de números forma el total mayor?
Haz un círculo alrededor de ellos.

Tarea de evaluación al final de la unidad

4. Resuelve cada problema. Usa dibujos, números o palabras para mostrar cómo obtuviste la respuesta.

Ayumi y Alejandro hornearon algunas galletas. Ayumi hizo 7 galletas y Alejandro hizo 8 galletas. ¿Cuántas galletas hicieron en total?

En el autobús había 13 niños. En la parada siguiente se bajaron 5 niños. ¿Cuántos niños quedaron en el autobús?

✓ Lista de énfasis matemáticos para *Formación de números*	Nombre del estudiante	Nombre del estudiante	Nombre del estudiante
1 Cuenta con exactitud un conjunto de hasta 45 objetos.			
2 Lee, escribe y pone números en secuencia hasta 100.			
3 Asocia palabras numéricas con los numerales escritos correspondientes.			
4 Usa números para anotar cuántos hay, en cantidades de hasta 40.			
5 Halla combinaciones de números hasta 15.			
6 Halla el total de dos cantidades hasta 20.			
7 Halla la mayor de dos cantidades, hasta 40 aproximadamente.			
8 Sabe las combinaciones de 10 (6 + 4, 8 + 2, etc.).			
9 Anota las estrategias para resolver problemas usando dibujos, números, palabras y ecuaciones.			
10 Hace, describe y extiende patrones de repetición, usando una variedad de materiales (ej.: acciones físicas, objetos, dibujos y números).			
11 Resuelve problemas en forma de cuento que requieren suma, usando: **a** modelado directo. **b** contar hacia adelante/contar hacia atrás. **c** relaciones numéricas.			
12 Resuelve problemas en forma de cuento que requieren resta, usando: **a** modelado directo. **b** contar hacia adelante/contar hacia atrás. **c** relaciones numéricas.			
13 Halla el total de varios números pequeños de un dígito.			
14 Halla diferentes combinaciones para un número (ej., 12 = 7 + 5, 8 + 4 y 3 + 9).			

Tareas de evaluación al final de la unidad

1A. Juguemos a "Adivina mi regla con botones".
Todos los botones que están en las cajas
son iguales de alguna manera. Dibuja cajas
alrededor de los demás botones que cumplan
mi regla.

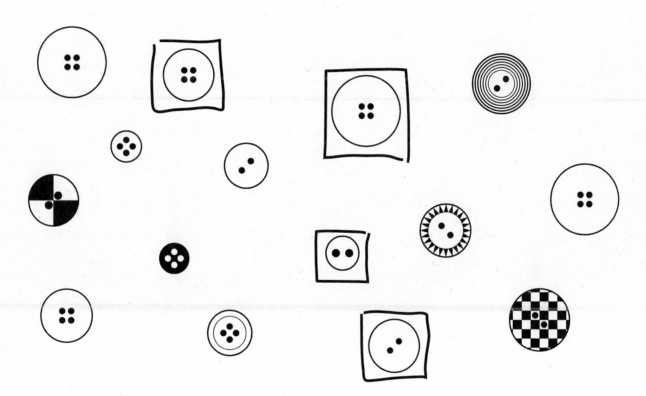

1B. ¿Cuál es mi regla?

81

Preguntas para las encuestas y reglas secretas

Tarea de evaluación al final de la unidad

2. Recorta las tarjetas con figuras de la Multicopia 14.
Clasifica las figuras en dos grupos.
Pega los dos grupos en esta página.
Nombra cada grupo.

Tarea de evaluación al final de la unidad

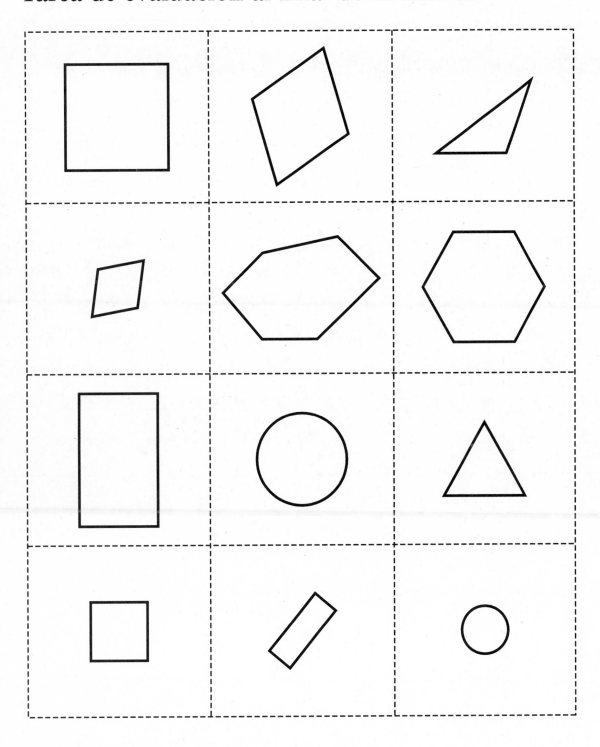

Tareas de evaluación al final de la unidad

3A. Esta encuesta se hizo en una clase de estudiantes de tercer grado.
Haz una representación de los datos en otra hoja.

Pregunta de la encuesta:
¿Prefieres comer helado en un **cono** o en una **taza**?

Alan ⬛	Shong ⬛	Paul ▽	Pamela ▽
Sara ▽	Masahiro ⬛	Joan ⬛	Susan ▽
Sean ▽	Linda ▽	Keri ▽	Kindra ⬛
Janet ⬛	John ⬛	Scott ⬛	Yuri ▽
Mark ▽	Niki ⬛	Allysa ▽	

3B. Cuenta un cuento sobre tu representación.
¿Qué aprendiste de ella?

_____

_____

_____

_____

☑ Lista de énfasis matemáticos para *Preguntas para las encuestas y reglas secretas*

	Nombre del estudiante	Nombre del estudiante	Nombre del estudiante
1 Identifica y describe atributos de varios materiales.			
2 Usa un atributo como base para clasificar y categorizar.			
3 Observa un conjunto de objetos agrupados e identifica el atributo común.			
4 Puede hacer un plan para reunir y anotar datos.			
5 Reúne y lleva la cuenta de datos.			
6 Clasifica y categoriza datos.			
7 Inventa y construye representaciones de datos.			
8 Explica e interpreta los resultados de encuestas.			
9 Entiende las representaciones de los demás estudiantes.			
10 Está familiarizado con el contenido de un calendario.			
11 Organiza y ordena datos sobre cumpleaños.			
12 Crea una representación de una línea cronológica.			
13 Puede crear una representación de datos que incluye varias categorías.			
14 Describe datos cuantitativa y cualitativamente.			
15 Interpreta conjuntos de datos que muestran valores y categorías al mismo tiempo.			
16 Compara dos conjuntos de datos.			

Tareas de evaluación al final de la unidad

1A. Usa bloques de patrón para llenar el
rompecabezas de esta página.
Luego anota las figuras que usaste,
pegando figuras de bloques de patrón
de papel con goma de pegar.
Anota el número de bloques que usaste.

Número de bloques: _____

Tareas de evaluación al final de la unidad

1B. Usa el mayor número de bloques que llene la figura._____

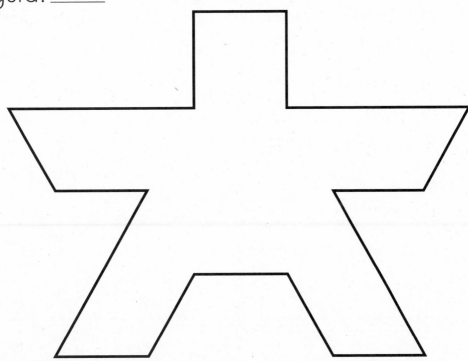

1C. Usa el menor número de bloques que llene la figura._____

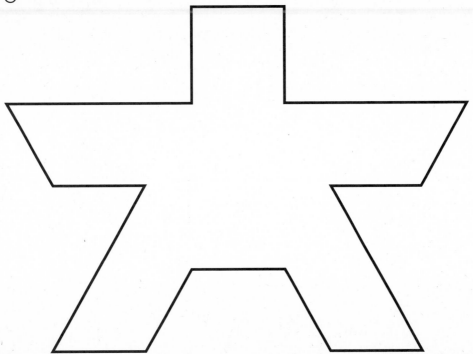

Tarea de evaluación al final de la unidad

2. Recorta las tarjetas con figuras de la Multicopia 20. Clasifícalas en dos o más grupos y pégalas en esta página. Nombra cada grupo.

Tarea de evaluación al final de la unidad

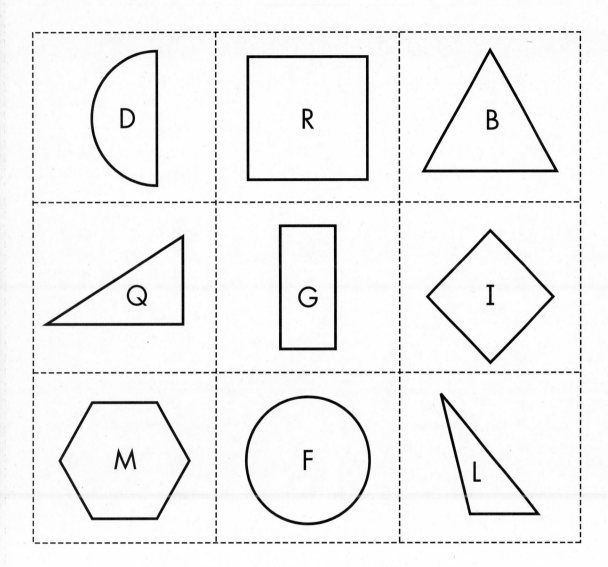

Tarea de evaluación al final de la unidad

3. Construye algo usando 3 ó 4 geobloques.
Dibuja lo que hiciste.

Tarea de evaluación al final de la unidad

4.   Usa las piezas que te dio tu maestro o
maestra para hacer una caja.

Lista de énfasis matemáticos para *Cuadrados para las colchas y pueblos hechos a base de cubos*

	Nombre del estudiante	Nombre del estudiante	Nombre del estudiante

Figuras bidimensionales

1	Observa, describe y compara figuras.			
2	Usa vocabulario matemático para describir y nombrar figuras.			
3	Describe características de los triángulos.			
4	Agrupa figuras de acuerdo con las características comunes.			
5	Compone y descompone figuras.			
6	Nota las relaciones entre figuras.			
7	Usa rotación y reflexión para organizar figuras.			
8	Llena con figuras una región dada.			
9	Visualiza y representa figuras.			
10	Está familiarizado con diversos cuadrados, rectángulos y triángulos.			

Figuras tridimensionales

11	Construye, observa, describe y compara figuras y objetos.			
12	Usa vocabulario matemático para describir figuras.			
13	Construye figuras tridimensionales a partir de figuras bidimensionales.			
14	Visualiza y describe prismas rectangulares.			
15	Compara y describe el tamaño, la forma y la orientación de los objetos.			
16	Une figuras tridimensionales para formar nuevas figuras.			
17	Crea y usa representaciones bidimensionales de lo tridimensional.			

Trayectos y patrones

18	Se fija en figuras en el entorno.			
19	Visualiza, describe y compara el trayecto entre dos puntos en el espacio y en una cuadrícula.			
20	Estima distancias.			
21	Visualiza y describe direcciones de rotación.			
22	Sigue, da y anota instrucciones de cómo moverse en el espacio y en un trayecto.			
23	Construye un patrón repitiendo una unidad cuadrada.			
24	Entiende cómo el hecho de cambiar la unidad afecta el patrón entero.			

Tarea de evaluación al final de la unidad

1. Tengo 13 creyones.
Algunos son azules y otros son rojos.
Tengo más creyones azules que rojos.
¿Cuántos creyones de cada color podría tener?

Tarea de evaluación al final de la unidad

2. Resuelve el problema.
Usa números, palabras o dibujos para
mostrar tu razonamiento.

En el parque hay 11 personas.
¿Cuántas piernas hay?

Tarea de evaluación al final de la unidad

3. Resuelve el problema.
Usa números, palabras o dibujos para
mostrar tu razonamiento.

En el parque había 13 patos y 9 gansos.
¿Cuántos animales había?

Tarea de evaluación al final de la unidad

4. Resuelve el problema.
Usa números, palabras o dibujos para
mostrar tu razonamiento.

En el lago había 18 patos.
9 se alejaron volando.
¿Cuántos patos quedaron?

✓ Lista de énfasis matemáticos para *Juegos de números y problemas*

	Nombre del estudiante	Nombre del estudiante	Nombre del estudiante
1 Lee, escribe y pone números en secuencia hasta 100.			
2 Halla el total de dos o más números de un dígito.			
3 Sabe las combinaciones de 10 (ej.: 6 + 4, 8 + 2 y 7 + 3).			
4 Está aumentando su familiaridad con pares de suma de un dígito.			
5 Halla combinaciones de números hasta 20.			
6 Entiende más, menos y lo mismo.			
7 Usa ecuaciones para describir la organización de objetos en grupos.			
8 Analiza imágenes visuales de cantidades.			
9 Anota soluciones con dibujos, números, palabras y ecuaciones.			
10 Halla más de una solución a problemas que tienen múltiples soluciones.			
11 Está desarrollando una estrategia para organizar conjuntos de objetos de manera que sea más fácil contarlos y combinarlos.			
12 Está desarrollando el sentido del significado de contar de 2 en 2.			
13 Se está familiarizando con los nombres de las monedas, sus valores y equivalencias.			
14 Identifica, describe, crea y extiende patrones de repetición.			
15 Representa un patrón de diferentes maneras (ej.: con acciones físicas, materiales concretos, dibujos y/o números).			
16 Identifica algunos patrones en la secuencia numérica y en la tabla de 100.			
17 Usa la calculadora como una herramienta matemática.			
18 Resuelve problemas en forma de cuento que requieren suma, usando: **a** modelado directo. **b** contar hacia adelante/contar hacia atrás. **c** relaciones numéricas.			
19 Resuelve problemas en forma de cuento que requieren resta, usando: **a** modelado directo. **b** contar hacia adelante/contar hacia atrás. **c** relaciones numéricas.			
20 Puede visualizar problemas en forma de cuento que requieran combinar cantidades de cambio desconocidas.			
21 Crea cuentos correspondientes a expresiones de suma.			

Tareas de evaluación al final de la unidad

Pon un objeto en un lado de la balanza.

1A. Pon algo **más pesado** en el otro lado.

Primer objeto:_____
Objeto más pesado:_____
Haz un dibujo para mostrar cómo queda la balanza.

1B. Pon algo **más liviano** en el otro lado.

Primer objeto:_____
Objeto más liviano: _____
Haz un dibujo para mostrar cómo queda la balanza.

1C. Pon algo que pese **lo mismo** en el otro lado.

Primer objeto:_____
Objeto del mismo peso:_____
Haz un dibujo para mostrar cómo queda la balanza.

Tareas de evaluación al final de la unidad

2A. Éste es un rompecabezas de bloques de patrón:

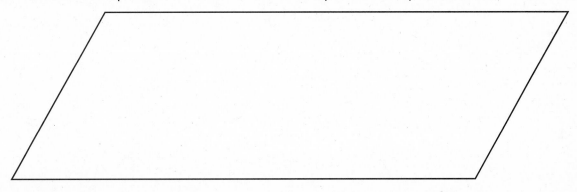

¿Qué conjunto de figuras de los de abajo llena el rompecabezas exactamente?

_____

Conjunto 1

Figuras	⬡	⏢	▱	▢	▱	△
¿Cuántas?	1	2	3	0	0	6

Conjunto 2

Figuras	⬡	⏢	▱	▢	▱	△
¿Cuántas?	2	1	1	0	0	3

2B. Escoge tu propio conjunto de figuras para llenar el rompecabezas. Anota cuántas figuras de cada tipo usaste.

Figuras	⬡	⏢	▱	▢	▱	△
¿Cuántas?						

Tarea de evaluación al final de la unidad

3. Escoge 2 recipientes. Escoge una herramienta de medición. Usa la herramienta para llenar cada recipiente con arena.

- ¿Cuánta arena entra en cada recipiente?
- ¿En qué recipiente entra más arena?

Escribe y dibuja acerca de cómo mediste.
Di lo que averiguaste.

Tarea de evaluación al final de la unidad

4. Escoge 5 cajas de la colección de "Cajas misteriosas". Mide con cubos el lado más largo de cada caja. Pon las cajas en orden de la más corta a la más larga. Muestra cómo organizaste las cajas. Incluye cuántos cubos de largo mide cada caja.

✓ Lista de énfasis matemáticos para *Más grande, más alto, más pesado, más pequeño*	Nombre del estudiante	Nombre del estudiante	Nombre del estudiante
Comprende lo que es el peso.			
1 Demuestra un sentido de *pesado* y *liviano* al tacto.			
2 Usa lenguaje para describir y comparar pesos.			
3 Usa una balanza para pesar objetos y compara el peso de diferentes objetos usando una balanza.			
4 Representa los resultados de las comparaciones de peso.			
Comprende lo que es la capacidad.			
5 Usa lenguaje para describir y comparar capacidades.			
6 Compara y mide capacidades usando unidades no estándar.			
7 Reúne datos, lleva la cuenta de los mismos y los interpreta.			
8 Estima el número de unidades necesarias para llenar un recipiente.			
9 Enumera y lleva la cuenta de cantidades hasta 50.			
10 Relaciona tamaño y forma con capacidad.			
11 Compara la capacidad de dos recipientes, llenándolos con sustancias continuas (como agua) u objetos discretos (como cubos).			
12 Compara la capacidad de más de dos recipientes.			
13 Llena con figuras un área dada.			
Comprende lo que es la longitud.			
14 Usa lenguaje para describir y comparar longitudes.			
15 Usa comparación directa de longitudes.			
16 Mide y compara longitudes usando unidades no estándar.			
17 Ordena longitudes.			
18 Representa mediciones con números, materiales concretos y/o dibujos, de una manera clara y ordenada.			
19 Desarrolla, describe y justifica técnicas para llenar recipientes, comparar capacidades y medir longitudes.			
20 Describe mediciones que no se pueden realizar en unidades enteras y exactas.			